Distillation 2024

Topical Conference at the 2024 AIChE Spring Meeting and 20th Global Congress on Process Safety

New Orleans, Louisiana, USA
24-28 March 2024

ISBN: 979-8-3313-1716-4

Printed from e-media with permission by:

Curran Associates, Inc.
57 Morehouse Lane
Red Hook, NY 12571

Some format issues inherent in the e-media version may also appear in this print version.

Printed with permission by Curran Associates, Inc. (2025)

For permission requests, please contact AIChE
at the address below.

AIChE
120 Wall Street, FL 23
New York, NY 10005-4020

Phone: (800) 242-4363
Fax: (203) 775-5177

www.aiche.org

Additional copies of this publication are available from:

Curran Associates, Inc.
57 Morehouse Lane
Red Hook, NY 12571 USA
Phone: 845-758-0400
Fax: 845-758-2633
Email: curran@proceedings.com
Web: www.proceedings.com

TABLE OF CONTENTS

DISTILLATION AND ABSORPTION PROCESSES FUNDAMENTALS, DEVELOPMENTS, AND APPLICATIONS II

DISTILLATION AND ABSORPTION PROCESSES FUNDAMENTALS, DEVELOPMENTS, AND APPLICATIONS III

DISTILLATION TOPICAL 2024: DISTILLATION ENERGY SAVING, PROCESS INNOVATIONS, AND PROCESS INTENSFICATION II

DISTILLATION TOPICAL 2024: DISTILLATION ENERGY SAVING, PROCESS INNOVATIONS, AND PROCESS INTENSFICATION III

DISTILLATION TOPICAL 2024: DISTILLATION TROUBLESHOOTING II

32a Quantifying the Effect of Stripping Factor on Distillation Efficiency Using Cyclopentanol/Cyclohexanol Test System

Rukhsar Ahmed, Tony Cai, Izak Nieuwoudt, Sayeed Mohammad, Clint Aichele

The optimal design of distillation columns requires reliable estimates of column efficiencies. Several models and correlations have been developed over the years to predict column efficiencies. However, these models may not be truly representative of actual efficiencies observed in the plant as most experimental data used to develop these models were obtained from distillation tests under total reflux conditions. Since commercial-scale distillation columns operate under non-total reflux, there is a need to obtain accurate experimental data on efficiencies under non-total reflux conditions. Such data will be helpful in developing reliable efficiency models based on both total reflux and non-total reflux conditions.

The stripping factor is an important variable that comes into the picture in the non-total reflux operation of a distillation column. The stripping factor is defined as the ratio of the slope of the equilibrium curve to that of the operating curve or line. In total reflux mode, the stripping factor is close to 1.0; however, in the actual commercial operation, the stripping factor deviates from unity. This study aims to fill the gap in the existing literature and provide new experimental data based on the non-total reflux mode of operation. Therefore, in this study, we present recent results obtained from an Oldershaw column under non-total reflux conditions for a highly viscous cyclopentanol/cyclohexanol test system. An Oldershaw column capable of operating in both total reflux and non-total reflux modes was used to measure point efficiency data over a range of conditions. The experimental results were also compared with the L/V ratios and the product concentrations obtained from column simulation in Aspen Plus®, both of which showed close agreement. Important findings from this ongoing study will be discussed

32b Packed Column Faq: Is There a Relationship between HETP and the Depth of Packing to Use in a Rate-Based Column Simulation?

Dale L. Embry, Hendrik A. Kooijman, Emma Russell

The use of non-equilibrium (NEQ) models (also widely known as Rate-Based models) for distillation has become increasingly common as such models can provide a more realistic model of real-world columns. We should note that two of the authors of this paper – RT and HK – are the creators of ChemSep, one of the more well-known rate-based column simulation tools; why we state this here will become evident in the next paragraph.

For tray-columns there is a one-to-one correspondence between the number of trays in a column and the number of stages used in a rate-based model of said column. For packed columns the situation is much less clear. For packed columns a single rate-based stage represents a slice of packing covering the entire column cross section (or, that part of the packing enclosed between the column shell and any internal dividing walls in a parallel column model of a DWC). The question most frequently asked of us (our FAQ) is a version of this one: "What depth of packing should we assign to a single rate-based stage?"

For a long time the standard suggestion has been to use three rate-based stages for each equilibrium stage with a depth of packing in each one equal to one-third of the expected HETP. We refer to this number as the Height Equivalence of a Rate-Based Stage (HERBS).

There is an immediate issue with this answer to this FAQ: Does stating that HERBS = HETP/3 not detract from the entire raison d'etre of a rate-based model? The answer to that is "undoubtedly yes." For many packings the expected HETP is known, but typically only for one or, at best, a small number of well-studied test systems. Does that expected HETP apply to all other cases? What about packings for which the HETP is not known? We acknowledge that the point of many rate-based simulations is to *predict* the expected HETP. Indeed, one should not have to know the expected HETP before conducting a rate-based simulation.

Given the uncertainty over the most appropriate value for the depth of packing to assign to a rate-based stage, we conducted an in-depth study to see if we can say something definitive about the value of HERBS to use in non-equilibrium column simulations of packed columns. What parameters affect the value of HERBS? How does packing size, packing type, column operational considerations (pressure and stripping factor) or simulation choices (mass transfer coefficient model, flow models) affect the answer. We also discuss the relationship between HETP and HERBS.

100b Dividing Wall Columns, a Step Towards *Net Zero* Emission

Manish Bhargava, Anju Patil

Dividing wall columns are known for energy reduction because they remove the inbuilt intrinsic thermodynamic inefficiencies as a result of which better product specifications are obtained. Also for getting desired product specifications they require lesser energy as compared to their conventional counterparts. A few revamps done at some of the known petrochemical complexes across the globe are sure to set an example to other facilities so as to take up challenges and identify area wherein this technology paves way to net zero.

Case Studies to be discussed:

- Debottlenecking of Hexane unit at South Asian refinery and its effect on reducing net emission from the facility
- Installation of Dividing wall DAS splitter in LOBS Unit of Indian refinery
- Xylene Recovery at South Asian refinery in one column system by use of Dividing wall column and its effect in reducing emissions.

175a Useful Rules of Thumb in Distillation

Henry Kister

Useful Rules of Thumb in Distillation

Henry Z. Kister, Fluor, Aliso Viejo, CA, USA

Rules of thumb should always be taken for what they are: rough criteria that reflect the experience of practitioners. They are not meant to replace detailed calculations such as those presented in most distillation texts. Their merit is in providing troubleshooters with a preliminary orientation while exploring potential trouble spots. Most of these rules of thumb apply to systems where there is a lot of experience. Many other systems are not covered by these rules.

This article presents common rules of thumb that have been useful to engineers engaged in troubleshooting distillation towers.

140a Structured Packing Technology for Post-Combustion Carbon Capture

Taylor Topham, Zack Bondley, Anand Vennavelli, Madi Lynch, Nathan Fine, Erik Meuleman

Post-combustion carbon capture is a vital technology in the future of energy as we strive for decarbonization. A new packing has been developed specifically to handle the conditions required for post-combustion carbon capture and provides an optimal combination of efficiency and low pressure drop. This packing and a proprietary solvent technology from ION Clean Energy were tested using a variety of point-source CO_2 compositions at Koch-Glitsch's R&D carbon capture facility. The testing shows clear improvements compared to traditional structured packing and solvent technologies to reduce the capital and energy requirements of full-scale carbon capture facilities.

This paper will present some of the results of the pilot plant testing and discuss potential applications.

140b Novel Modeling of Liquid-Phase Hydrodynamics and Mass Transfer in Structured Packings

Luke Macfarlan, R. Bruce Eldridge

Structured packings are advanced column internals for chemical separation equipment, including distillation columns. This type of packing offers high separation efficiency, low pressure drop, and high throughputs. Opportunity for further improvement to these packings exist, but development efforts are challenged by an incomplete understanding of the hydrodynamics and mass transfer inside the system. This presentation describes a novel computational fluid dynamics (CFD) approach to capture the hydrodynamics and mass transfer in these packings. Multiple validation sources confirmed the accuracy of the simulations. CFD predictions of the liquid holdup were compared with two separate experimental data sources and showed strong agreement. Predictions of the liquid flow angle were compared with an analytical expression and showed exceptional agreement. The CFD predictions of the Fanning friction factor aligned with falling film theory. Lastly, the predicted liquid mass transfer coefficients showed strong agreement with experimental data taken from a pilot-scale column. The validated CFD model subsequently supported a packing design study, where the packing design was modified, and the hydrodynamic and mass transfer performances were characterized. Reducing the channel opening angle to 60° and the channel inclination angle to 30° increased the liquid mass transfer coefficient by 12.4% and 18.0%, respectively. This improvement in mass transfer was driven by higher liquid flow angles and, therefore, higher liquid upheaval.

140c RMXR-Ring, RVT Process Equipment´s 4th Generation Random Packing Development and Performance

Jose Sola, Benjamin Pechtold, Christian Geipel

The development of RVT´s 4th generation metal random packing family, RMXR-Ring, will be presented using the standard 2-inch nominal size RMXR#2 as an example. Initially used development steps included 3D printing and filling tests of prototypes as well as simulations of randomly bulked packing beds based on discrete element method (DEM). These results were used in a next step to estimate the packing pressure drop and capacity level via CFD simulation. The derived finally developed geometry of the RMXR #2-Ring was later implemented in a cut/bending production tool for its mass production and its hydraulic performance and mass transfer efficiency characterized with pilot plant columns and common standard test systems [1]. The results of these performance studies will be presented and compared to common 2nd and 3rd generation metal rings.

The results demonstrate that the developed RMXR geometry significantly surpasses the hydraulic capacity of standard 3rd generation rings by achieving a significantly lower pressure drop while delivering an outstanding mass transfer efficiency level both for absorption and distillation applications.

9a A Next Generation Moving Valve Device

Neil Sandford, Anand Vennavelli, Charles Griesel, Emily Ruhl

The FLEXIPRO valve tray provides high capacity, excellent separation efficiency and a wide operating range up to 5:1, which is well into the range traditionally associated with moving valve trays. This presentation will introduce a new floating version of the FLEXIPRO valve which provides the same high tray efficiency across an even wider operating range. We will showcase how the floating FLEXIPRO valve tray is an easy drop-in replacement for all existing, earlier generations of moving valves. The performance will be demonstrated including capacity and efficiency data from pilot plant testing and illustrated with example of commercial usage cases.

9b C2 Splitter Revamp with HiFi Plus Trays Part 1

Andre Bernard, Senthil Krishnamoorthy

Construction of a new polyethylene facility by NOVA Chemicals in the Sarnia, Canada area has driven the need for expanding the existing NOVA Corunna's ethylene cracker capacity by about 50%. This expansion project was completed during the 2022 plant turnaround and involved significant scope on the 2 C2 Splitter towers operated at the site. The primary tower was partially re-trayed and equipped with an additional side reboiler. The second tower was fully re-trayed from 2 pass conventional trays to HiFi high-capacity trays. A side reboiler was added along with a second reboiler and overhead condenser. This paper discussed the design, installation, and operating strategy.

9c C2 Splitter Revamp with HiFi Plus Trays Part 2

Senthil Krishnamoorthy, Andre Bernard

Nova Chem Sarnia in Sarnia, ON wanted to increase their Ethylene Plant capacity to feed their new polyethylene facility. There are two C2 Splitters in this plant. This paper focus mainly on the secondary C2 Splitter. This C2 Splitter was equipped with conventional 2-pass trays. To meet the target capacity, it was found that this column was not able to handle the required loads. Based on previous experience with primary C2 Splitter, NovaChem decided to revamp this column into Shell HiFi Plus Trays with few additional heat exchangers. Nova Chem and Sulzer worked together to implement the revamp and HiFi trays were installed in fall 2022. Then the plant started up and has been in operation for the past several months. This paper will compare operating data before and after revamp and it will also discuss the impact on capacity and efficiency of Shell HiFi Plus trays over 2-pass trays.

9d Increased Tower Run Length with Anti-Fouling Trays.

Mark Pilling, Siang Hua Lee, Peng Cheng Hou

This presentation will discuss the various Sulzer anti-fouling tray technologies. It will describe successful revamps of a column prone to fouling and serving the purification of MDI, an aromatic di-isocynate that is used as monomer in the manufacture of polyurethane. Thanks to the implementation of high-performance anti-fouling tower internals, the operating run length was increased 1.5 times after the first revamp and then doubled after the second revamp.

52a Design of Robust, Energy Efficient, Packed (Dividing-Wall) Column Line-Ups

Hendrik A. Kooijman, Emma Russell

It has long been known that Dividing Wall Columns (DWCs) can provide chemical process plants with more energy efficient separation trains while simultaneously reducing the capital cost. Until now, however, there existed no easy-to-use complete design process that made it straightforward to find the best, balanced, and robust DWC configurations. In this paper we outline a step-by-step methodology that leads to robust distillation column designs for regular as well as dividing wall applications with packing. Our approach considers many factors normally not part in conceptual design but that *do* influence actual hardware design and operation:

1. Thermal (in)stability
2. Variations in the feed composition (and state)
3. Pressure drop
4. Maldistribution
5. Marangoni effects
6. Liquid de-wetting
7. Foaming
8. Proper calculation of the total annualized cost (TAC) of DWCs*

(* TAC calculation models in the literature invariably omit many factors that impact the true cost. These include unequal effective column section heights either side of a wall, empty space, flow (re) distributors, column swaging, and sloped walls.)

For DWCs we will show that wall position and bed configurations can easily be optimized by simple, fast and efficient parametric variations of the stage efficiency and liquid splits. We illustrate our comprehensive approach by considering several industrial processes. In this paper we briefly outline our parallel column model that undergirds the proposed methodology. Then we discuss important extensions that make possible this comprehensive design process.

MACHINE-LEARNING BASED PRESSURE DROP PREDICTION FOR STRUCTURED COLUMN PACKINGS

AIChE PROCEEDINGS

Marcus Schlager[1,2], *Tossapon Katongtun*[2], *Sanphawat Phromphithak*[2], *Nattawut Khuenkaeo*[2], *Nakorn Tippayawong*[2], *Markus Lehner*[1]

[1] *Montanuniversität Leoben, Leoben, Austria*
[2] *Chiang Mai University, Chiang Mai, Thailand*

Introduction

Accurate prediction of gas-side pressure drop in packed distillation or absorption columns is essential for apparatus dimensioning. As a results, many semi-empirical models such as Stichlmaier et al. (1989), Billet and Schultes (1991) or Maćkowiak (2010) have been developed. The main drawback of these modeling approaches is that they typically rely on one or more experimentally determined packing-specific parameters that must be obtained through measurements before being able to describe a packing.

An alternative to semi-empirical physics-based models is the use of data-driven modeling methods. Data-driven modeling refers to the creation of predictive models directly from data, without specifying explicit theoretical assumptions or physical relationships between model features and the target variable(s). To derive such models, regression algorithms from the domain of machine learning are commonly used. In addition to yielding predictions, these models may further be used to assess the influence of model features on the target variable, thereby enhancing the understanding of the examined system.

The objective of this research is to derive a regression model for the prediction of structured column packings' pressure drop that fully eliminates the need for packing-specific model parameters. Once such a model is trained, packing pressure drop can be predicted directly from packing descriptors. In particular, the proposed model can be used to calculate dry and irrigated pressure drop of novel packing geometries, without first having to characterize such geometries experimentally. Two independent approaches are explored in this work: For the "meta-model" approach, the form factor of the Maćkowiak model serves as target variable for a regression model. For the "direct approach", pressure drop is directly modeled without further specifying relationships between the variables, thereby following the data-driven paradigm.

Theory/calculation

The Maćkowiak pressure drop model

Calculation of dry pressure drop using the model of Maćkowiak is based on the Ergun-equation:

$$\frac{\Delta p}{H} = \psi \frac{(1-\varepsilon)}{\varepsilon^3} \frac{F^2}{d_\mathrm{P} K}$$

where ε is packing porosity and gas load is given in the form of the F-factor ($F = \sqrt{\varrho_\mathrm{G}} u_\mathrm{G}$). The wall-factor assumes a value of $K = 1$ for structured packings and the "particle diameter" d_P is calculated as follows:

$$d_\mathrm{P} = 6 \frac{(1-\varepsilon)}{a_\mathrm{geo}}$$

The resistance coefficient ψ is a function of gas phase Reynolds number $\mathrm{Re_G}$:

$$\psi = \left[\frac{725.6}{\mathrm{Re_G}} + 3.203\right](1-\varphi)$$

$$\mathrm{Re_G} = \frac{K}{(1-\varepsilon)} \frac{d_\mathrm{P} u_\mathrm{G} \varrho_\mathrm{G}}{\eta_\mathrm{G}}$$

The packing-specific form factor φ is used to adjust model predictions to experimental observations. For this, φ is determined through pressure drop measurements and may assume values between 0 and 1.

Considered model features

The following packing properties are chosen as model features: packing porosity ε, specific geometric area a_geo, packing angle α, channel side length s, channel angle β, free surface area a_free and packing surface finishing (binary variable IsWavy). An illustration of geometric packing properties is given in Figure 1. A total of 7 packing properties are considered as model features.

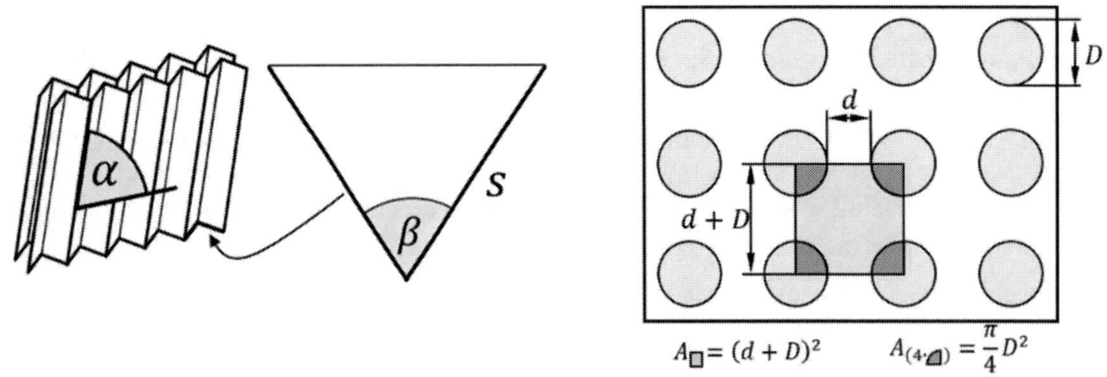

Figure 1. Model features to describe structured column packings. Left: Packing segment (packing angle α). Middle: Individual packing channel (channel angle β, channel side length s). Right: Detail of packing surface with equally spaced perforations.

Metal packings are further categorized into two distinct groups regarding packing surface. The first group consists of column packings where parallel ridges and grooves are embossed into the packing surface ("wavy" surface). The second group of packings has not been embossed ("smooth" surface). Whether a packing has smooth or wavy surface can easily be assessed visually. The two cases are illustrated in Figure 2. Examined packings manufactured from plastic or ceramic lack an embossed surface and are therefore assigned to the group of packings with "smooth" surface. Although some surface patterning may be present in the case of individual plastic packings, it is significantly less pronounced than the embossed ridges present in "wavy" metal packings.

Regression models

A total of 4 regression models are assessed for their suitability of capturing the influence of packing properties on the packing form factor and pressure drop. Examined modeling techniques include linear regression, Random Forest, Gradient Boosting, and a Multi-Layer Perceptron. The models were fitted using the sklearn library (Pedregosa et al., 2011) for the Python programming language.

Linear regression (LR) models the relationship between input variables and the model target by fitting a linear equation via the method of least squares. LR is simple to implement and easy to interpret. The influence of each model feature is represented by a model coefficient that indicates strength and direction of the change in the target variable. As a downside, LR assumes only linear relationships between model features and the model target and is therefore unable to capture more complex relationships. For this work, LR serves as a baseline model, ensuring that the increased complexity of the other models translates into improved model predictions.

Random Forest (RF) is an ensemble learning approach that combines several randomized decision trees and aggregates their predictions. Each decision tree is trained on only a random subset of the data and only considers a random subset of model features. Random Forest is versatile, has shown excellent performance in many settings and returns measures of variable importance for better model interpretability (Biau and Scornet, 2016).

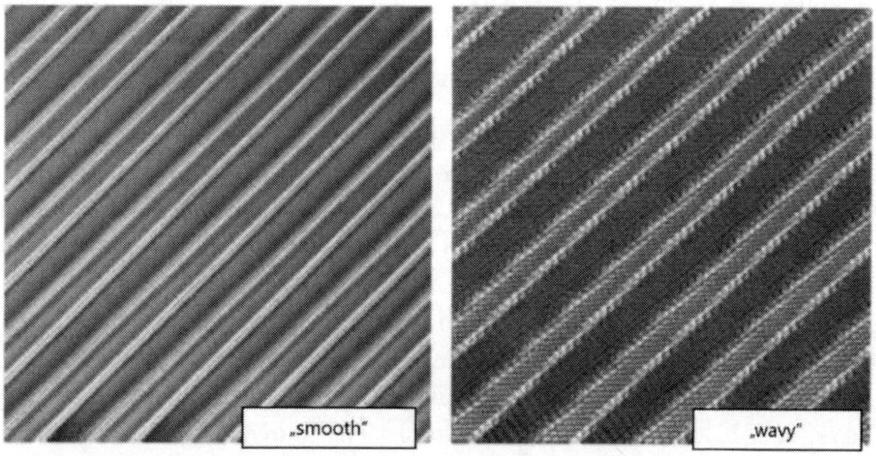

Figure 2. Two metal packings with different surface finishing. Left: "Smooth" surface. Right: "wavy" surface (embossed surface). Neither column packing shown is perforated.

Gradient Boosting (GB) builds an ensemble of decision trees sequentially, where each subsequent tree is selected so that it minimizes the prediction error of the previous tree.

A multi-layer perceptron (MLP) is a type of artificial neural network composed of multiple layers of nodes arranged in a feedforward manner. The model consists of an input layer, one or more hidden layers, and an output layer. Each neuron in a layer is connected to every neuron in the subsequent layer, and each of these connections is associated with a weight. MLP performance depends heavily on its architecture (number of hidden layers, choice of activation functions, etc.). For this analysis, an MLP model with 3 hidden layers containing 10 nodes each is considered.

Model cross-validation

Model performance is assessed through K-fold cross-validation. For this, the dataset is first shuffled and then randomly split into K subsets, using $K - 1$ subsets for training the model and the remaining subset for model validation. This process is repeated K times, with each subset used once as the validation data. This procedure generates exactly one corresponding model prediction for each data point, the comparison of which is used to evaluate model performance. In this work $K = 5$ is set.

In the case of the meta-model approach, leave-one-out cross-validation (LOOCV) is used for model validation due to the relatively small total number of data points (form factor for 34 packings). LOOCV is a special case of K-fold cross-validation where K equals the number of instances in the dataset. For each iteration, one data point is held out as the validation set, and the model is trained on the remaining K-1 data points.

Materials and Methods

Experimental apparatus

The experimental apparatus for packing characterization (Figure 3) consists of a saturator column (600 mm) and a measurement column (422 mm) manufactured from polypropylene. All hydraulics measurements are conducted using the air/water system. A detailed description of the apparatus is given in Schlager (2022).

The apparatus operates as follows: First, ambient air is introduced into the saturator column where it is contacted with deionized water to ensure constant experimental conditions. After exiting the saturator column through a demister mesh, gas phase relative humidity and temperature are determined. Relative humidity is over 90% for all experiments. Next, the gas phase flows into the measurement column via a tube-type gas distributor and enters the column packing under investigation at the bottom side and flows vertically upwards through the entire column packing (maximum packing height up to 2 m). Gas phase pressure drop over the investigated packing is measured using a differential pressure gauge with measuring points above and below the column packing. The blower is located downstream the measurement column. Minimum gas flow is approx. $F = 0.5\ \mathrm{Pa}^{0.5}$ while a maximum gas flow of approx. $F = 5.0\ \mathrm{Pa}^{0.5}$ can be attained. However, depending on the type of packing installed, maximum gas flow may be significantly lower during experimental runs.

For pressure drop determination, liquid load in the measurement column is first set to a constant value between 10 and 100 m³/(m²h). Next, the lowest gas load of approx. $F = 0.5\ Pa^{0.5}$ is set. After reaching stationary conditions, pressure drop is sampled for multiple minutes to average out any minor fluctuations. Gas load is then incrementally increased to repeat the pressure drop measurement procedure at multiple gas loads. A measurement series is concluded after maximum gas load is reached or after packing flooding. The measurements are repeated at different liquid loads.

Investigated column packings

In total, data from 34 structured column packings is used in this study, an overview of which is provided in Table 1. Although most of the examined column packings are commercially available products, some were specially manufactured for this study. For example, the data set includes packings with packing angles of up to 80°. As a result, the validity range of the derived model is increased. Packing properties are as follows: Porosity is between 0.72 and 0.99 (mean value 0.94), specific geometric area is between 64 and 500 m²/m³ (mean value 235 m²/m³), packing angle is between 37 and 80° (mean value 53°), channel side length is between 0.8 and 7 cm (mean value 2.4 cm), channel angle is between 76 and 113° (mean value 92°) and free surface area is between 0 and 0.43 (mean value 0.1). 25 (73.5%) of packings have perforations. 21 packings (61.8%) are made of metal, 7 (20.6%) are made of plastic and the remaining 6 (17.6%) are made of ceramic. Within the group of metal packings, packings are further differentiated according to surface finishing: 10 packings (29.4%) have a smooth surface, 24 packings (70.6%) have wavy surface.

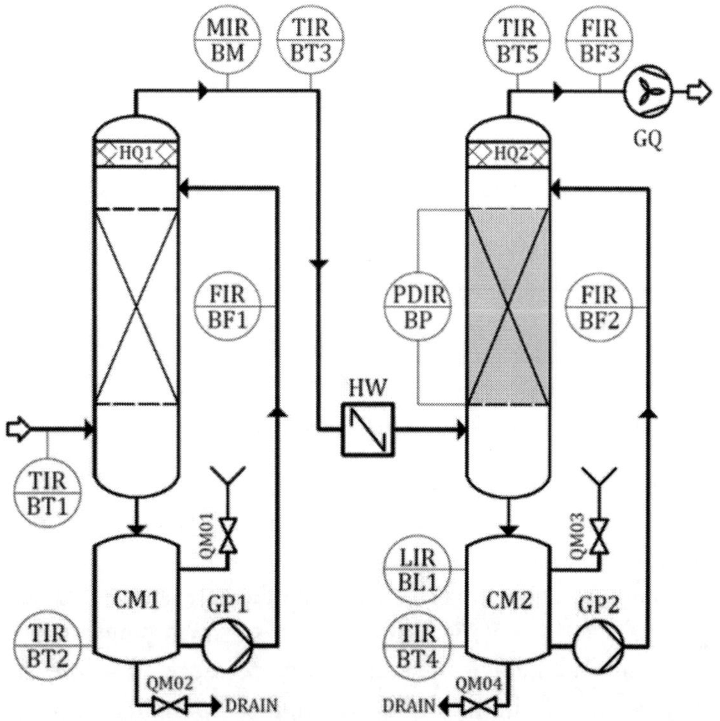

Figure 3. Experimental apparatus for column packing hydraulics characterization. Left: Saturator (inner diameter 600 mm). Right: Measurement column (inner diameter 422 mm).

Table 1. Properties of the investigated column packings.

Packing	ε m³/m³	a_{geo} m²/m³	α °	s cm	β °	a_{free} m²/m²	IsWavy	φ
Metal packings								
Mellagrid 64Y	0.99	64	45	7.0	81	0.000	0	0.75
RMP N 125X	0.99	128	60	3.2	77	0.102	1	0.93
RMP N 125Y	0.99	128	45	3.2	77	0.102	1	0.79
RMP N 200X	0.99	202	60	2.1	111	0.102	1	0.95
RMP N 200Y	0.99	202	45	2.1	111	0.102	1	0.87
RMP N 250X	0.99	252	60	1.6	87	0.102	1	0.92
RMP N 250Y	0.99	252	45	1.6	87	0.102	1	0.85
RMP N 350X	0.98	355	60	1.1	86	0.102	1	0.93
RMP N 350Y	0.98	355	45	1.1	86	0.102	1	0.84
Mellapak 250Y smooth	0.99	252	45	1.6	87	0.102	0	0.85
Mellapak 500Y	0.98	500	45	0.8	76	0.102	1	0.84
RMP SP 250-1	0.99	250	53	1.7	99	0.080	1	0.89
RMP SP 250-2	0.99	250	61	1.7	99	0.080	1	0.92
RMP N 250 W80	0.99	250	80	1.6	77	0.080	1	0.97
RMP N 250 W70	0.99	250	70	1.6	85	0.080	1	0.95
RMP N 250 W37	0.99	250	37	1.6	89	0.074	1	0.70
RMP N 250Y gl. oL.	0.99	250	45	2.0	92	0.000	0	0.78
RMP N 250Y gl. mL.	0.99	250	45	1.7	103	0.074	0	0.80
RMP N 250Y we. oL.	0.99	250	45	2.0	101	0.000	1	0.79
RMP N 250Y-1 we. mL.	0.99	250	45	1.8	93	0.094	1	0.82
RMP N 250Y-2 we. mL.	0.99	250	45	1.8	95	0.090	1	0.81
Plastic packings								
RPP N 125X	0.95	125	60	5.1	89	0.216	0	0.93
RPP N 250Y	0.90	250	45	1.8	96	0.201	0	0.82
RPP SP 250	0.90	250	59	1.9	102	0.201	0	0.86
RPP S 250	0.90	250	68	2.0	105	0.201	0	0.88
Flexipak 3X	0.98	125	60	4.9	95	0.433	0	0.91
MP N 125.X-PP	0.95	125	60	5.2	92	0.302	0	0.93
Plastic packing, Chinese origin	0.95	125	60	5.1	113	0.221	0	0.88
Ceramic packings								
K125X	0.86	125	60	2.7	88	0.000	0	0.84
K125Y	0.88	125	45	3.2	92	0.000	0	0.51
K250X	0.80	250	60	1.7	92	0.000	0	0.89
K250Y	0.80	250	45	1.6	92	0.000	0	0.78
K350Y	0.78	350	45	1.2	94	0.000	0	0.80
K450X	0.72	450	60	1.0	91	0.000	0	0.95

Results

Experimental determination of the packing form factor

To evaluate the form factors of the examined column packing, the model of Maćkowiak is fitted to each packing's dry pressure drop measurement data by varying φ until best fit (least squares method). Experimentally obtained values for φ are given in Table 1. For the packings investigated in this work, the form factor assumes values between $\varphi = 0.51$ (K125Y) and $\varphi = 0.97$ (RMP N 250 W80). The mean form factor is $\varphi = 0.85$. In summary, the model Maćkowiak is found to reproduce dry pressure drop measurement results with excellent accuracy ($R^2 = 1.0$) in the entire investigated operating range.

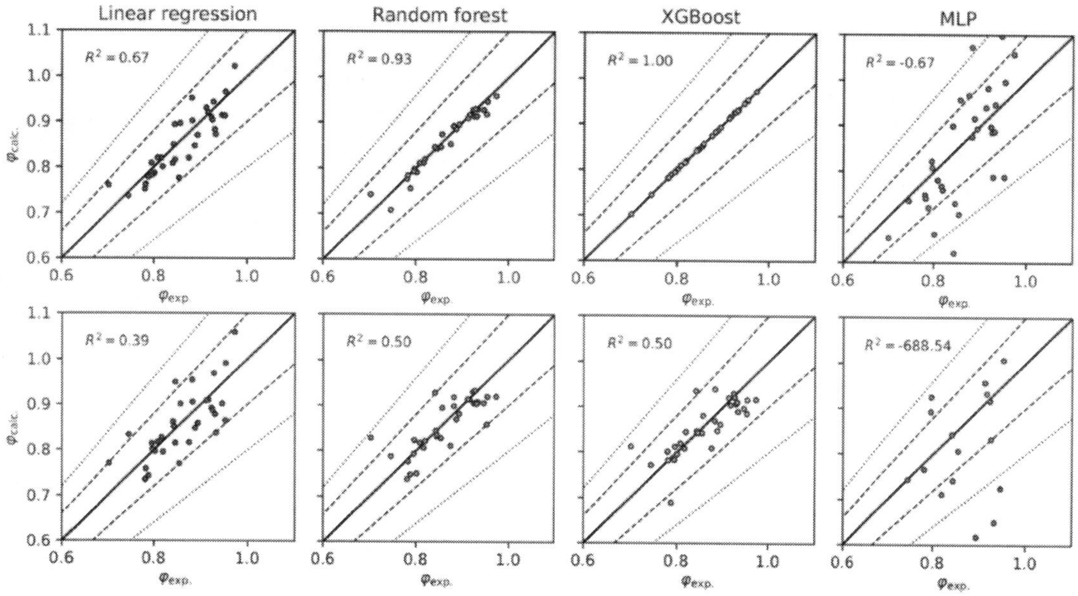

Figure 4. Modeling results for the packing form factor φ (meta-model approach). Top row: model fit results. Bottom row: 5-fold cross-validation results. Dashed lines: ± 10 %. Dotted lines: ± 20 %.

Meta-model approach

For the meta-model approach, the regression models were fitted to the packing form factor φ. Model predictions and cross-validation results are given in Figure 4.

Even using the simple linear regression model, good agreement ($R^2 = 0.67$) between model predictions and experimental data is achieved. All model predictions lie within 10% to experimental results. Using RF, very good ($R^2 = 0.93$) agreement between model predictions and measurement data is achieved. The maximum deviation between an experimental value and the corresponding model prediction is about 6%. GB can fit form factors with excellent accuracy ($R^2 = 1.0$) and there is practically no difference between model predictions and experimental values. MLP model predictions are unsuitable as variance of model predictions is greater than the total variance in the form factor. This illustrates that the relatively high number of model parameters of the MLP cannot be meaningfully determined by the comparatively low number of available data points. Taken together, RF and GB perform best for form factor modeling in the meta-model approach.

Regarding cross-validation, predictions for all models are less accurate on test data that differs from the training data. The best results are achieved by RF and GB (both $R^2 = 0.5$). Remarkably, even though GB can fit form factors with excellent accuracy it performs just as well in cross-validation as RF. In summary, none of the examined models perform exceptionally well during cross-validation. A major reason for this lies the models having to extrapolate from the training data in some cases. For example, if the packing with the lowest porosity is used as test data, it is necessary for the model to extrapolate to a lower porosity based on the remaining training data. However, the decision-tree based models (RF, GB) and the MLP generally fail to extrapolate. R^2-values given in the bottom row of Figure 4 can therefore be regarded as a conservative estimate when generating model predictions that lie within the range of validity.

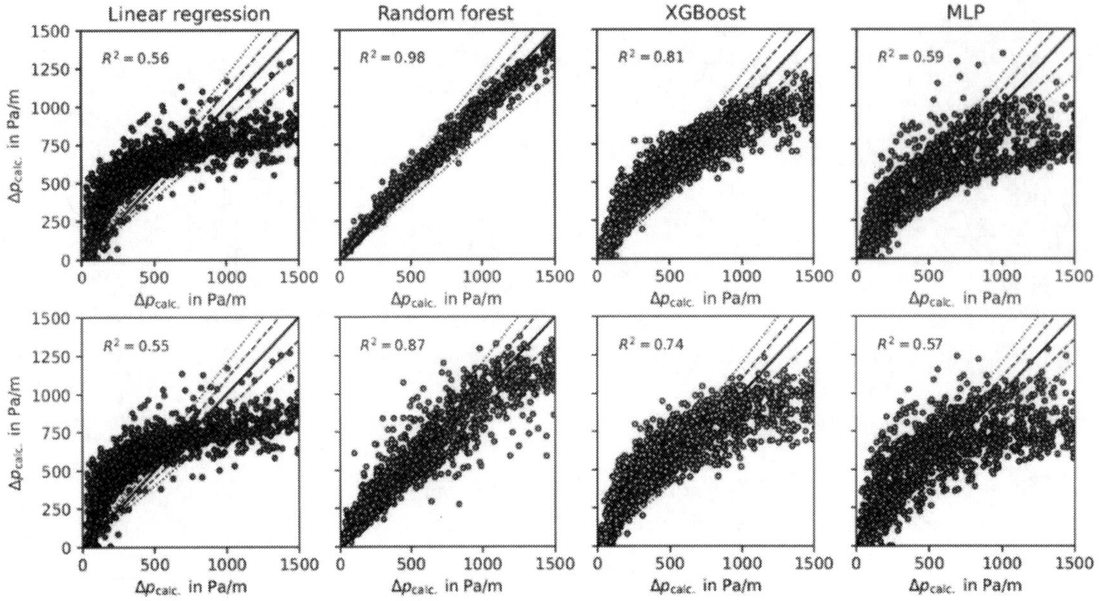

Figure 5. Modeling results for packing pressure drop Δp (direct approach). Top row: model fit results. Bottom row: 5-fold cross-validation results. Dashed lines: ± 10 %. Dotted lines: ± 20 %.

Direct approach

For the direct modeling approach, the regression models were fitted directly to pressure drop measurement data consisting of a total of 1457 individual measurements. Model predictions and cross-validation results are given in Figure 5.

The most striking observation is the excellent ($R^2 = 0.98$) fit of the RF regression model. Most model predictions lie within 10% of experimental values and only single model predictions deviate more than 20% from measurements. Model results are in good agreement with experimental data in the entire examined range, although the model slightly underpredicts pressure drop at higher pressure drops starting at about 1000 Pa/m. The GB model is also in good agreement with experiments, but this is mostly the case for lower pressure drops as starting at about 1000 Pa/m the model increasingly underpredicts experimental values. MLP modeling results follow the same general shape, but scatter is increased so that MLP predictions are found worse than GB. LR only poorly fits to pressure drop data. LR first strongly overpredicts experimental data and later strongly underpredicts measurements for experimental pressure drop over about 750 Pa/m.

Regarding cross-validation, the RF model reaches a very good score of $R^2 = 0.87$. The model gives good predictions for pressure drop over the entire examined range. Although GB gives good prediction accuracy ($R^2 = 0.74$), model predictions for higher pressure drop are again in poor agreement with measurements. MLP in its current form delivers cross-validation results only slightly better ($R^2 = 0.57$) than linear regression ($R^2 = 0.55$).

The currently poorer performance of the GB and MLP models may be explained due to the lack of extensive hyperparameter-tuning in this first analysis. GB and MLP models are particularly dependent on model hyperparameters.

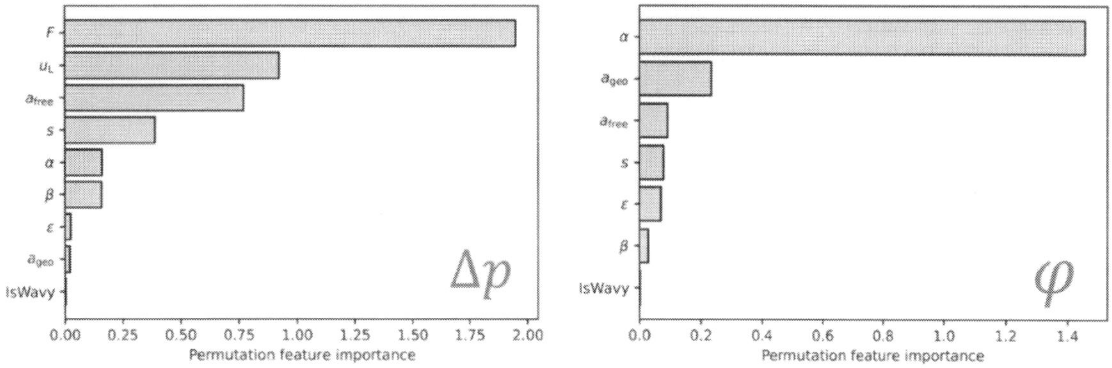

Figure 6. Permutation feature importance (Random Forest regression). Left: Influence of operating point (F, u_L) and packing properties on pressure drop. Right: Influence of packing properties on packing form factor.

Feature importance analysis (Random Forest regression)

Permutation feature importance (PFI) of the fitted random forest model m is calculated from the tabular dataset D using the following procedure: First, the reference R^2 score s of model m is calculated for data D. Next, column j of D corresponding to model feature j is randomly shuffled to create a corrupted version of the data $\widetilde{D}_{k,j}$. From this, score $s_{k,j}$ is calculated. This is repeated a total of K times. PFI i_j of feature j is then evaluated using:

$$i_j = s - \frac{1}{K}\sum_{k=1}^{K} s_{k,j}$$

The calculation procedure is then repeated for each model feature j.

PFI calculation results for the random forest regression model are given in Figure 6. The results for direct modeling of pressure drop suggest that the column's operating point is central in the description of pressure drop, since model features F and u_L are found to have the highest PFI. In the operating range investigated, gas load exerts a greater influence on pressure drop than the liquid load. Model features describing packing properties play a smaller role in explaining pressure drop compared to the operating point. Among these features a_{free}, s and α have the highest PFI scores.

However, care must be taken when ranking PFI scores due to multicollinearity issues in the dataset, meaning that some of the model features are also correlated among each other. For example, strong $(R^2 = -0.82)$ correlation is found between features s and a_{geo}. Although these two features are in principle independent of each other, the correlation likely stems from the manufacturing process of the packings. For packings with large a_{geo}, s must be small to conveniently stack many layers closely together. Multicollinearity generally leads to lower PFI scores of correlated features, as the regression model can still gauge a feature's influence on the target variable even after data corruption during the PFI calculation procedure.

An interesting perspective emerges when looking at the PFI scores for the meta-model approach. Since Mackowiak's model is a semi-empirical model, the effect of numerous packing parameters on pressure drop is already represented through physics-based model equations. The empirical model parameter φ then can be said to contain additional physical relationships that are not explicitly represented in the model equations. PFI of the meta-model approach may thus provide information on how φ is composed.

The analysis shows that the single most important packing property influencing φ is the packing angle α. This is not at all a surprising result, as α is not considered in Mackowiak's model equation. The model feature with the second largest PFI score is geometric packing area a_{geo}. This would suggest that – although included – a_{geo} is not represented fully accurately in Mackowiak's equations. Minor influences on φ include a_{free}, s, ε and β. Again, caution must be applied in interpretating these results due to multicollinearity.

Conclusions and Outlook

The objective of this research is the development of a regression model for structured packing pressure drop that is independent of packing-specific parameters. For this, two independent modeling approaches were explored: In a meta-model approach, the form factor of the Maćkowiak model served as target variable for a regression model. For the direct approach, pressure drop is directly modeled from packing properties. Four regression models including linear regression, Random Forest, Gradient Boost, and a Multi-Layer Perceptron were examined.

Random Forest regression is found to yield strong results in model fit and cross-validation score. In the direct modeling approach, Random Forest regression results in excellent model fit ($R^2 = 0.98$) while most model predictions are within 10% to experimental results. The cross-validation score of $R^2 = 0.87$ suggests good generalizability of model predictions to other packing than the ones examined in this work. Taken together, the Random Forest regression model is already a serious contender of established semi-empirical modeling approaches without even requiring packing-specific constants.

The GB and MLP regression models in their current form require more thorough hyperparameter-tuning. Due to the already strong performance of the RF model, it will be interesting to see whether the increased complexity of the GB or MLP models can translate into improved model predictions even after hyperparameter-tuning.

Further, the presented methods may lead to a better understanding or further developments of semi-empirical models, as these models can assist in elucidating the influence of packing properties on the empirical parameters of current semi-empirical models. In this work, the influence of packing properties on the packing form factor of the Maćkowiak model was briefly discussed.

Next steps in this research include the further improvement of the RF, GB and MLP regression models and the consideration of liquid-holdup. If successful, the result is a packing hydraulics model that does not rely on packing-specific parameters.

Acknowledgement

This work has received funding through the research project "SME 5.0 – A Strategic Roadmap Towards the Next Level of Intelligent, Sustainable and Human-Centred SMEs" funded in the European Union's Horizon MSCA 2021 programme under the Marie Skłodowska-Curie grant agreement No. 10108648.

References

1. Stichlmair, J., Bravo, J.L., Fair, J.R. (1989), "General model for prediction of pressure drop and capacity of countercurrent gas/liquid packed columns," *Gas Separation & Purification 3.*
2. Billet, R., Schultes, M. (1991), "Modelling of pressure drop in packed columns," *Chemical Engineering & Technology 14.*
3. Maćkowiak, J. (2010), *"Fluid dynamics of packed columns,"* Springer Berlin Heidelberg.
4. Pedregosa, F., Varoquaux, G., Gramfort, A. et al. (2011), "Scikit-learn: Machine Learning in Python," *Journal of Machine Learning Research 12.*
5. Biau, G., Scornet, E. (2016), "A random forest guided tour," *TEST 25.*
6. Schlager, M., Haushofer, G., Triebl, A., Wolf-Zöllner, V., Lehner, M. (2022), „Determination of volumetric mass transfer coefficients of structured packings at two different column diameters," *Chemical Engineering Research and Design 186.*

52c On the Inability of Weeping Correlations to Predict the Behavior at Zero Liquid Load: New Experimental Data for Trays, Correlation Development, and Dynamic Weeping

Hannes Raddant, Patricia Polgár, Markus Illner, Christian Hoffmann, Verena Fath, Philipp Hamberger, Laura Fender, Julia Kirschbaum, Bernd Metzen, Jens-Uwe Repke

Distillation columns and especially tray columns are still among the most important and widely used thermal separation units in chemical industry. In most chemical production facilities, they are operated at steady-state and ensure, for example, the purification of valuable reaction products at high throughput. The actual design and sizing of distillation columns adheres to specific boundary conditions given by the production process or plant (e.g. separation task, production capacity, operational flexibility). Due to today's improved understanding fluid dynamics in distillation columns and advancements in process modeling and simulation, occurring phenomena within tray columns and their feasible operating window are often sufficiently predictable as long as the plant is operated as intended. If the required operation window is left, especially in short time horizons, safety-relevant column conditions could occur. Normally, these operation conditions appear during malfunctions result in dynamic processes inside distillation columns especially regarding fluid dynamics and relevant safety scenarios take place outside well described operation regions of distillation columns. To prevent bursting of the column and release of hazardous substances, it is hence required to install safety equipment, whose selection and sizing is based on worst case predictions (e.g. total reboiler failure) and resulting mass and heat flows. It is thus desirable to investigate the dynamic weeping behavior for such a scenario and generate suitable correlations and model structures to predict the expected re-evaporation and the resulting pressure increase more accurately. This way, a better understanding of process dynamics inside tray distillation columns can be obtained via experiment-based correlation development and a safer plant operation can be realized through better understanding of the process as well as dynamic simulations. In addition, investment cost for safety equipment can be reduced by avoiding oversizing.

Knowledge or correlations describing the fluid dynamics for these scenarios are primarily available for rather simple tray configurations in the open literature. The existing literature contains a large quantity of sieve tray data, as for example Lockett et. al investigated experimentally the weeping inside a sieve tray column and developed a weeping correlation based on their steady-state experimental data [1]. Wijn developed a weeping model, with the focus on the description of the clear liquid height under low gas loads, for trays and validated it on available test data [2]. However, the applied gas load factors described as hole load factor in Wijn only varied on the range of 0.09-0.90 m/s and thus relevant shutoff scenarios are not fully captured. Especially for gas loads < 0.20 m/s and high liquid loads > 14 m^3/m^2h, the qualitative weeping behavior for trays is insufficiently described. Experiments of safety-relevant scenarios inside of distillation columns need to be carried out in order to develop correlations for a better understanding of the plant behavior in the non-designated and safety-relevant operation window. It is furthermore desirable to have a model-based description of the occurring fluid dynamic phenomena in distillation columns outside their operational limits, such as very low or vanishing gas load factors and resulting weeping behavior, available. Therefore, this contribution discusses a typical distillation safety scenario for the separation of wide boiling multicomponent mixtures using tray columns and highlights experimental investigations to enable the development of suitable weeping correlations for trays. Combined with a relatively high liquid holdup on the trays compared to structured packings, an evaporator malfunction leads to dynamic weeping of large quantities of the liquid hold-up on the trays and in the downcomers. Coming from steady-state operation, especially liquid rich on light boiling components is weeping from the upper section of the column into the column sump or reboiler resulting in a potentially very fast evaporation of the light boiler. As a consequence, the pressure inside of the column can increase rapidly and strongly, due to re-evaporation.

Steady-state weeping experiments were carried out for weeping correlation development and to improve the process knowledge regarding trays. Therefore, our experimental setup consists of a DN 300 column made of glass and holds a mounting structure for exchangeable trays, liquid downcomers, as well as exchangeable weirs. A liquid cycle is established using a gear pump, feeding liquid from a separate collector tank to the top of the column. The liquid is then collected after passing the column and transported back to the storage tank, whereby a lower collector tray ensures separate flow measurements of liquid weeping through the trays and liquid exiting via the downcomer using magnetic-inductive flow meters. The gas load is applied by using a speed-controlled fan and gas distributor at the bottom of the column, while also providing the option for abrupt shutoff of the gas flow. The setup is fully automatized using ABB Freelance and operable in a range of 0.0-1.0 $Pa^{0.5}$ and 1.3-35.5 m^3/m^2h for gas load and liquid load, respectively. Weeping rates from 0 up to 100 % as well as tray pressure drop are reliably tracked. As a test system, water/air is used at ambient temperature and pressure conditions (both measured). Experimental results shown in this contribution focus on the application of standard trays for two different weir heights.

The experimental results from our own investigations for trays are presented and discussed. We used our data to develop a new weeping correlation, which captures the weeping behavior over a large range of gas load factors as well as liquid loads. Therein, special focus is laid on the mathematical structure to ensure usability within dynamic tray column models and robust convergence behavior. Finally, the structure and applicability of this new correlation will be critically discussed in comparison the gained experimental data and validation runs, as well as applicability for safety assessments and dynamic process modelling.

Based on our experiments, we present a comparison of the measured weeping rates with existing literature correlations. These results show a different quantitative and qualitative weeping behavior for specific trays. Most importantly, no weeping experiments are published so far which are able to correctly capture the behavior in the vicinity of zero weir load. This inability of the published data respectively correlations make them virtually inapplicable to the safety-relevant scenarios. In addition, dynamic weeping experiments for different initial conditions will be presented.

[1] Lockett, M. J., and S. Banik. 'Weeping from sieve trays.' Industrial & Engineering Chemistry Process Design and Development 25.2 (1986): 561-569.

[2] E.F. Wijn, On the lower operating range of sieve and valve trays, Chemical Engineering Journal, Volume 70, Issue 2, 1998, Pages 143-155, ISSN 1385-8947, https://doi.org/10.1016/S0923-0467(98)00089-X.

79a Investigation of Crystallization Fouling on Column Internals

Korbinian Inderwies, Harald Klein, Sebastian Rehfeldt

Fouling is the formation and deposition of solids on surfaces in mass and heat transfer applications. Due to the impairment of equipment, such as columns, fouling is a highly undesirable process in industry.

Fouling represents one of the main causes of malfunctions in distillation columns. However, only a few studies on fouling in columns are available so far. To characterize the fouling tolerance of different column internals, a test rig has been set up and put into operation at the Institute of Plant and Process Engineering at the Technical University of Munich. A column with a diameter of 150 mm is used to investigate random and structured packings, and a column with a diameter of 300 mm is used for the investigation of multiple trays.

Since fouling is mostly a slow, gradual process, which can last for years without malfunctions, an acceleration of the experiments is necessary to generate results in a reasonable time. Crystallization fouling is a suitable mechanism for these investigations since the mechanism can be accelerated, and the crystallization fouling is responsible for a significant number of malfunctions in columns in the industry.

The crystallization fouling is forced in the test rig using a saturated sodium chloride water solution and air. Due to evaporation in the column, the solution gets supersaturated and causes the salt to crystallize, which then partly deposits on the internals. The fouling on the internals narrows the free cross-sectional area for the gas flow, which leads to an increase in the pressure drop over the column.

Since the different internals have different separation efficiencies, the fouling tolerance of the internals is difficult to compare directly. To consider the different separation efficiencies, the pressure drop increase is related to the evaporated water mass, which acts as a measure of the actual mass transfer. The fouling tolerance of the internals can be compared according to the amount of evaporated water until a specific pressure drop is reached. In our investigations, this limit is 1.5 times the initial pressure drop. The more water is evaporated, the higher the fouling tolerance of the internals. In addition, the pressure drop can be measured over several levels in the column so that a local detection of the fouling is possible.

These investigations can quantify the fouling tolerance of different internals and operating conditions to identify the most suitable trays and packings for processes prone to crystallization fouling. In this contribution, the test rig and first measurement results will be presented.

79b Digital Twin Distillation Operations Training

Michael Taube, Isuru A. Udugama, Brent Young

The safe and on-specification control of distillation columns is one of the challenging aspects of operating a modern-day chemical production facility. In comparison to the previous decades, today many distillation columns are operated at tighter product specifications and at higher product and process recovery rates. And as we transition to more renewables, operation during increasing variability in energy inputs due to demand response and supply upsets is increasingly likely. As a result, even experienced process operators and engineers need to think about the complex interactions between thermodynamics, internal hydraulics, mass and energy balance limitations and controllers when making changes. Considering these shifts and the inevitable loss of experienced operators and engineers there is a need for continuing education and competence building to be carried out in the domain of distillation process control.

Traditional continuing education and operator trainers even when built from dynamic process simulators are often based on local dynamic models based on simple linear but abstract transfer function models or resistance capacitance network models. These local models are useful for simulation of relatively mild process upsets or more carefully curated upsets such as start-up and shutdown that are pre-programmed. We propose and advocate the use of dynamic digital twins based on a combination of mechanistic and data driven modelling for a wider range of operational upsets, e.g., such as recovery from significant and rapid power turndown for demand response. In this presentation we describe the key concepts and advantages of our real-time approach to operations training simulation over the aforementioned traditional approaches with the use of a validated dynamic digital model of a real methanol refining column.

79c On the Improper Implementation of the Murphree and Vaporization Efficiencies in Process Simulators

Holt Ragsdale, Zachary Braswell, Brian Hanley

Many of the commonly used chemical process simulation tools today offer the capability to apply efficiency adjustments to individual equilibrium stages or sections, aiming to better match the separation performance of those stages with real-world tray performance. Some simulators even permit the use of component efficiencies at the stage level or within stage sections. However, we believe that two prevalent efficiency models, namely vaporization efficiency and Murphree vapor phase efficiency, have often been implemented incorrectly in these simulators. The resulting errors in the column temperature, composition, and liquid/vapor phase profiles can be profound. The errors arise from three significant conceptual inconsistencies:

1) The assumption that liquid and vapor streams always leave a stage at the same temperature, even when considering stage or component efficiencies.

2) Mandating the specification of efficiencies for all 'C' components when only 'C-1' are truly independent.

3) In the case of vaporization efficiency, assuming that equal component efficiencies for all components is mathematically and thermodynamically consistent.

These three assumptions can result in unfeasible column composition and temperature profiles - the result of serious violations of fundamental thermodynamic principles. To rectify these mathematical and thermodynamic issues, we propose the following adjustments:

1) Always treating the liquid and vapor streams leaving a stage as bubble point liquids and dew point vapors.

2) Specifying efficiencies for only 'C-1' components.

We have created a rigorous column model in Excel using the Aspen Properties V12 Excel add-in that treats vaporization and Murphree vapor efficiencies properly. We will compare the results of several simulations performed with this corrected model to results from a number of commercial simulators available today.

79d Challenges and Opportunities in Large-Scale Distillation Research

Tony Cai, Izak Nieuwoudt

Distillation is widely recognized as the cornerstone of mass transfer processes due to its remarkable versatility and extensive application in the chemical and process industries. In distillation column design, the accurate prediction of liquid/vapor distribution, tray/packing hydraulics, scale-up effect, and other factors is pivotal, as they profoundly influence the performance of column internals. Nevertheless, the study of these phenomena faces many limitations in small-scale laboratory columns, primarily due to the substantial disparities in the 'shape of a theoretical stage' between laboratory and industrial-scale distillation research. Fractionation Research, Inc (FRI) has been at the forefront of industrial-scale distillation research since 1952. This presentation will delve into the significance of industrial-scale distillation studies, shedding light on the challenges and complexities that accompany such research endeavors. Moreover, it will outline the measures and solutions devised to tackle these challenges and capitalize on the opportunities presented in large-scale distillation studies. Furthermore, we will discuss the significance and influence of research in industrial-scale distillation on energy efficiency and sustainability.

119a Systematic Workflow to Design Electrified Distillation Systems for Binary and Ternary Separations

Akash Nogaja, Mohit Tawarmalani, Rakesh Agrawal

In contemporary chemical process development, there is a notable surge of interest in the electrification of unit operations. This transition is a direct response to the Paris Agreement and the rising proportion of renewable energy sources in the electricity mix (Energy Information Administration, 2021). Distillation, as a dominant technology in downstream processes, stands out as a prime candidate for electrification. Enhancing lower-grade heat, such as condenser heat operating at lower temperatures, through the application of electricity in reboiler heating represents a highly efficient approach to electrify a distillation system, and heat pump technology embodies this innovation (Null, 1976).

Vapor recompression, bottom flashing, and external loop heat pumps, which are the primary alternatives, have been thoroughly examined in the existing literature and have showcased remarkable success in various industrial applications. Conversely, tools such as feed preheating, intermediate reboilers/condensers, and multi-effect distillation have demonstrated significant efficiency improvements under specific feed conditions (Tumbalam Gooty et al., 2021). The integration of these tools with heat pumps has the potential to significantly reduce electricity consumption even further. Therefore, the primary goal of this current research is to develop a decision framework that can guide industrial practitioners when considering the electrification of distillation technology.

In this research, we embark on a two-fold approach. First, we illustrate various electrification alternatives using both single and dual compressors. Subsequently, we construct shortcut models for estimating multi-component phase change temperatures (Nogaja et al., 2022), compressor work consumption, to evaluate the performance of all the different alternatives. These shortcut models facilitate a rapid initial assessment to identify configurations suitable for more comprehensive simulation. Finally, we develop an easy to use distillation framework relying only on the feed properties, while producing high purity products, to guide the design of optimal heat pump assisted distillation (HPAD) system.

References:

Energy Information Administration, U. (2021). *Annual Energy Outlook 2021 Narrative*. www.eia.gov

Nogaja, A. S., Mathew, T. J., Tawarmalani, M., & Agrawal, R. (2022). Identifying Heat-Integrated Energy-Efficient Multicomponent Distillation Configurations. *Ind. & Eng. Chem. Res.*, *61*(37).

Null, H. R. (1976). Heat Pumps in Distillation. *Chem. Eng. Prog.*, *58*, 58–64.

Tumbalam Gooty, R., Chavez Velasco, J. A., & Agrawal, R. (2021). Methods to assess numerous distillation schemes for binary mixtures. *Chemical Engineering Research and Design*, *172*, 1–20.

119b Economic and Energy Usage Comparison of Conventional and Dividing Wall Distillation Columns

Natalie Czarnecki, R. Bruce Eldridge

Residue curve maps (RCM) were used to determine the products of direct conventional, indirect conventional, and middle wall dividing wall column (DWC) distillation. This process was repeated for all 125 RCM present in Perry's Chemical Engineering Handbook. Categories of RCM were developed based on an azeotrope being present in one product stream or no azeotropes being present in any product stream. Model chemical systems, with open literature available vapor-liquid equilibrium data, were chosen for select RCMs that fell into each category. The theoretical separations were validated using Aspen Plus simulations. All simulations were performed using Siemens' HEEDS optimization software. The optimized simulations of direct, indirect, and middle wall DWC were used to size and cost the process equipment. The comparison of the optimized simulations illustrates the economic savings in both fixed capital investment and utilities costs due to the use of a middle wall DWC.

119c Process Intensification for Diethyl Carbonate Production By Combining Reaction and Separation

Donggun Kim, Minyong Lee, Yongbeom Shin, Jeongwoo Lee, Jae Lee

In this study, we propose an improved process for the production of diethyl carbonate from ethylene carbonate and ethanol through transesterification reactions in a reactive distillation process, aiming to enhance energy efficiency and economic feasibility. The feedstock for diethyl carbonate production, dimethyl carbonate, has been challenging to separate due to the formation of a low-boiling-point azeotrope with methanol. Special distillation techniques such as pressure swing distillation or extractive distillation have been required for its separation. To address this issue, our study integrates the production processes of dimethyl carbonate and diethyl carbonate using only reactive distillation, eliminating the need for separating the azeotrope. Two design alternatives are proposed depending on configuration of reactive distillation. The first alternative utilizes two reactive distillation columns, allowing the direct transfer of the product from the first column to the second without separating the azeotrope. This reduces the overall number of required equipment and results in a 12% reduction in total utility consumption and total annual cost compared to the existing process. The second alternative conducts all transesterification reactions in a single reactive distillation column, achieving a 24% reduction in total utility consumption and a 14% reduction in total annual cost. These newly proposed processes not only enhance economic feasibility but also reduce carbon dioxide emissions by 15% and 11%, respectively.

156a Process Intensification in Post-Combustion Carbon Capture

Ming Yang Lee

Post-combustion carbon capture via solvent absorption has been executed in several industrial-scale projects in recent years. However, prohibitive capital and operating costs have hindered the realization of many projects under development. Furthermore, there are concerns over the emissions of solvents and other pollutants, particularly for carbon capture plants built near residential areas. While process licensors bear a big responsibility to resolve these issues with enhanced solvents and process configurations, better mass transfer technologies and column designs can also contribute to this process intensification activity.

The duty of the flue gas blower upstream of the CO_2 absorber contributes the third highest operating expense in the carbon capture plant, after the CO_2 stripper reboiler and the CO_2 compressor system. Minimizing the CO_2 absorber column pressure drop helps to reduce the blower duty, and can be accomplished by the MellapakCC^{TM} structured packing, which offers lower packing pressure drop without sacrificing CO_2 absorption efficiency.

Solvent emissions are often reduced to sub-ppm levels by using two pumparounds in the water wash section of the CO_2 absorber. A more efficient column design could be achieved by replacing one pumparound with a once-through design, which firstly saves one set of pumps and heat exchangers, and secondly leads to near-zero emissions. The once-through design results in ultra-low specific liquid loads, requiring the use of a specialized structured packing with an ultra-low liquid load distributor.

Experimental results from both pilot and industrial columns will be presented to validate the efficiency and hydraulic performance of the mass transfer technologies discussed.

156b Innovation in Spray Technology- Ultra Low Entrainment Sprays for Vdu Wash Beds

Alejandro Lago

Excessive entrainment of liquid gas oil droplets in the wash section of a vacuum distillation column can lead to critical process reliability issues and distillate purity issues. Extensive droplet testing shows that entrainment can be reduced by using a new spray nozzle technology designed to generate coarser diameter droplets, ensuring a larger percentage of wash oil makes it to its intended target of wash-bed packing. Extensive testing shows that a new spray nozzle technology (Lechler SMDmax^{TM}) can produce a much more favorable spectrum of droplets in comparison to traditional maximum free passage style spray nozzles. This favorable spectrum of droplets can be correlated to a significant reduction in entrainment from a spray header in counter current flow. This presentation focuses on the use of this new spray nozzle technology in distillation applications; however, these process benefits can also be realized in many other entrainment-sensitive spray applications.

156c Energy Saving Separation System Methodologies

Izak Nieuwoudt

The purification of chemical compounds is key to so many consumer products that determine our standard of living and quality of life. These purification steps could be energy intensive. Electrification has been touted as a means of reducing the fossil-based energy consumption of these processes. There are other ways of reducing the energy consumption. In this presentation several case studies will be discussed to show how alternative separation technologies can significantly reduce energy consumption. Coupling these technologies with electrification and other CO_2 reduction strategies could potentially present more economical alternatives than the one-size-fits-all strategies that are being considered.

156d Measuring Cells As a Design Tool for Columns with Structured Packings.

René Pöschmann, Georg Brösigke, Ilja Ausner, Ansor Gaebler, Jens-Uwe Repke

Established methods for the design of packed columns for distillation, absorption and desorption are based on simplified and extensive model assumptions. Due to the high complexity of the phase distribution within the column and insufficient knowledge of the mass transfer behavior of real multicomponent systems, predictions on the actual separation performance of industrial columns can bear a high level of uncertainty.

A widely used parameter for characterizing the separation performance of structured packings is the HETP value, which, among other things, depends strongly on the physical properties of the components to be separated. Therefore, in many cases basic tests in calibrated lab columns are required for an unknown non-ideal system to separate. Scale-up of the plant to production scale is then done using the HETP value determined with standard test systems that are usually chosen from the list of recommended test systems for distillation columns compiled by Onken and Arlt[1].

The disadvantage of this column-based methodology is the comparatively large amount of investigated system required, which often cannot be provided and the high costs for conducting the experiments. For a fast and cost-efficient determination of the HETP value, respectively a weakening factor, which translates the HETP from an ideal to a real system, an apparatus as small as possible while maintaining scale-up capabilities is therefore desirable.

Pöschmann et al.[2] developed a measuring cell to fulfill these requirements and provided the proof of concept through experiments with the recommended test system cyclohexane – n-heptane (C6-C7) by comparing the results of the separation performance as well as the dry and wet pressure drop to columns of diameters ranging from DN 50 up to DN 250.

It was found, that on the one hand the measuring cell, with a channel of unperforated packing sheets, showed good agreement with separation performance as it would be expected for a perforated Mellapak™ 500.Y. While on the other hand, the DN 250 column showed a poorer performance with an unperforated packing compared to the perforated one. The hypothesis was, due to the relatively short packing bed and the missing perforation of the packing used within their experiments, the phase maldistribution caused the deterioration in separation performance for the bigger columns. To verify this hypothesis, it is necessary to eliminate one of the described effects. Therefore, further experiments within the DN 250 column were conducted with the same unperforated packing, but with a packing bed height of 2.5 m. To eliminate any effects of the used test system, the new experiments were conducted with the recommended test system usually used by Sulzer, Chlorobenzene - Ethylbenzene (CB-EB).

As the measuring cell showed consistent results for the system C6-C7 and good transferability to the expected performance of the perforated Mellapak 500.Y, experiments with CB-EB were also conducted in the measuring cell to compare these to the data of the DN 250 column.

The measuring cell shows good agreement with the expected performance of the Mellapak 500.Y (perforated) but the DN 250 column with an unperforated packing still underperforms for the new setup with CB-EB. Thus, the perforation of the packing can be identified as an important parameter for reducing the phase maldistribution. Missing perforation negatively effects the mass transfer efficiency. Furthermore, the results suggest that the measuring cell represents an ideally wetted channel of perforated packing sheets even though the packing installed within the measuring cell was not perforated.

To review this even further, a rate-based model with the mass transfer model from Olujic[3] was set up to describe the experiments. The model by Olujic was developed for the prediction of mass transfer in perforated packings. The modeling results show good agreement to data the measuring cell (unperforated), whereas the data from the DN 250 column (unperforated) are not matched. To examine the possible limits of the measuring cell, an unperforated Mellapak 250.Y packing was installed and tested with C6-C7 and CB-EB, where the measuring cell could also show consistent results as well as good alignment with the expected performance of a perforated Mellapak 250.Y.

All the findings described support the hypothesis, that the perforation of the packing is important for the liquid distribution and therefore the mass transfer efficiency. Furthermore, the findings indicate that the measuring cell represents an ideally wetted channel of perforated packing sheets and therefore has great potential as a tool in the design process of columns with structured packings for the separation of non-ideal mixtures.

The authors gratefully acknowledge financial support from the German Federal Ministry of Education and Research (BMBF) and project supervision by the DLR Project Management Agency for Environment and Sustainability for the project 'ReProvAP' (Reduzierung der klimarelevanten Prozessemissionen durch die verbesserte Auslegung von strukturierten Packungskolonnen) with the funding codes 01LJ2002F.

[1] Onken, U., Arlt, W., 1990. Recommended Test Mixtures for Distillation Columns. The Institution of Chemical Engineers ISBN 0852952481

[2] Pöschmann et al., 2023. Will laboratory and pilot plant columns soon become superfluous? – A concept for the determination of structured packing characteristics in a measuring cell under distillation conditions. Sep.Purif. Technol. 325, 124617, DOI: 10.1016/j.seppur.2023.124617

[3] Olujic et al., 2004. Predicting the Efficiency of Corrugated Sheet Structured Packings with Large Specific Surface Area.; Chem. Biochem. Eng. Q. 18 (2) 89–96.

186b Increase Throughput with Improved Product Specifications Using Dividing Wall Technology

Manish Bhargava, Amit Kanda

Dividing wall technology is increasingly used in the industry to produce more throughput in existing columns.

This presentation explores the revamp journey of two commonly used separation sequences:

Case study 1: A sequence of Naphtha Splitters were revamped multiple times over its 15 years of life cycle. The original configuration of columns was equipped with conventional sieve trays. In the 1st revamp the column internals were changed to high capacity packing. Though the capacity increase was more than 40%, the product specifications were compromised.

In the 2nd revamp, an additional side cut was introduced to improve product specifications. The drawback was a reduction of top and bottom product yields.

Finally, the column was revamped to a dividing wall column. **The revamp provided a capacity increase of an additional 40% over that achieved by high capacity internals**. There was simultaneous improvement in the product specifications and yields.

Case Study 2: This case study involved an indirect sequence of a depropanizer followed by propylene splitter producing chemical grade propylene. The two columns were equipped with high capacity trays. Besides a 30% increase in through put, DWC revamp provided higher propylene recovery and lower RVP of bottoms C4 product used for gasoline blending.

AUTHOR INDEX